Sphereology Part 1: Up to the Third Dimension

By Bransford C.

To the Reader:

This book contains the beginnings of mathematic principals inspired by physics and the universe, primarily gravity and orbiting particles. Beyond this point are the basic ideas of this math proposal along with how and why I came to the concepts I did. Please note that the concepts here are new and theoretical and there *may* be some flaws. Though if there is enough interest in these concepts as a mathematics, it could be used to accurately portray graphs that better fit this universe of ours.

Note: at the end of this book is a format summary and description of everything discussed until then.

Chapter1: The Pros and Cons

We have an amazing mathematics system that is thousands of years in the making. Every time a problem comes up, mathematicians worldwide work until an agreed solution is found. After all this time, these great minds have mastered the standard Cartesian grid that has set the basis for everything we do in a daily life. Yet one detail gives a constant impression that the math we use doesn't completely define the universe, as math should. This one small detail is at the limits of our number system, and is defined as unreachable.

I grew up with a talent for math, and sincerely enjoyed problem solving. The idea that made this an easy study was math was all one collective pattern that fit together perfectly. It wasn't until middle school that I had the first clue that there was something incomplete with math, when a calculator divided anything by zero it just read error. I asked my math teachers about it and their response was always that it is impossible. I couldn't get the idea out of my head, it was math, there had to be a solution. The idea continued to haunt me every now and then, and every time I could never think of a solution. It wasn't until I happened to watch a show on black holes a few months later that an idea came. What stood out was when one of the scientists described the singularity of the black hole having matter in no space, making it *infinitely* dense. That was the answer, infinity! A moment of relief came over for having the answer in front of me, and not only that but an answer that exists. This moment soon diminished at another thought: why did the calculator read error, and why did my math teachers say it is *impossible*. I continued to watch the show with more interest. One scientist described that 1/0 was just a value in math we could easily sway from, though in physics is a real monster.

A year or two later, with many side-projects in between, I had a realization. Any axis that makes up a grid is a line with two sides that went to infinity. With reasoning, wouldn't a line be the same as a line segment connecting -∞ and ∞.

This idea sparked the concept that ∞ could be a limit, not only an unreachable value. Though no one could have guessed at this point where this first concept would lead.

Chapter2: Limits to Infinity

An important step to a revolutionary idea is to master it before making it public (so to minimize competition). The best starting point is a linear graph, luckily I only needed two examples to make a clear point. Below are two graphs on a grid; (y=2x) and (y= -∞/2) that go to the limits of the grid.

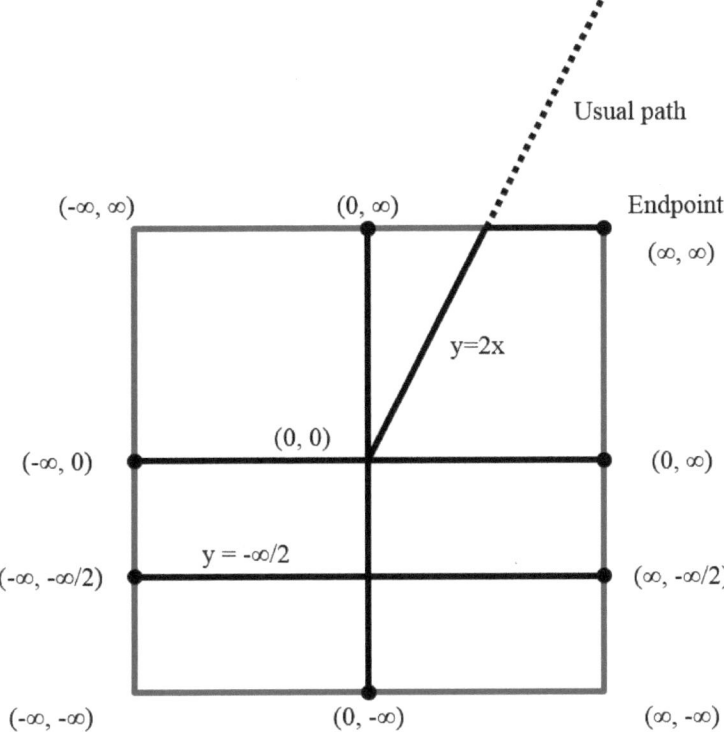

Imagine you're sitting still in the middle of a square room with walls, and there is absolutely NO friction in this experiment you are participating in. You're allowed one pushing force to get you around the room. Let's say you're afraid of bouncing in unpredictable ways, so you ask them to push you in a perpendicular path into the wall. They do so and when you meet the wall instead of bouncing off, the wall completely absorbs the impact and you're left against the edge of the room. This is the example of the (y = -∞/2) graph, there's only one axial correlation, and when it reaches the limit, that is the endpoint. Since the people doing this test are giving you another chance you want to try a different method. You think that something similar will happen if you head straight to the corner, so you ask to be pushed into a wall at an angle. They place you at the room's center again and give you a little push, once again you slide across the floor. This time when you meet the wall you find you're still sliding along the edge of the room, and you keep going until you reach the corner, then you completely stop. This time there were two correlations of direction much like the (y=2x) graph. When the graph met the limit it would usually keep going in the same path, though can't due to the limit; the alternative is to follow at least one of the correlations until both correlations are at their limits.

After understanding linear graphs well enough, the next step I believed should be exponents beginning with a simple (y=x²) graph. The major concept that needed to be figured out was when the graph *first* meets the y-limit; the limit of this graph as x goes to infinity is just infinity. After plugging ∞ in for y, the equation to find x is x=$\sqrt{y}$, which in this case becomes x=$\sqrt{\infty}$. This is still ∞ since $\sqrt{\infty} = \sqrt{1/0} = 1/0 = \infty$. This led me to believe ∞ acted as if it were one, which isn't possible if you look at another way to graph (y=x²). The problem is after translating the equation into the ordered pair $(\sqrt{\frac{1}{y}} * \frac{y}{1}$, y) {If you don't believe this works test it yourself with any number and this equation}. Solving for x when y=∞ gives us $\frac{1}{\infty} * \frac{\infty}{1}$, usually we could just say this gives us one, however if we translate this to $\frac{0}{1} * \frac{1}{0}$ then we're left with $\frac{0}{0}$. Then the puzzle begins: does $\frac{0}{0}$ leave us with 0, 1, or ∞?

Chapter3: The Dawn of a New Mathematics

The exponent problem taunted me for months, especially when a percent came up. How could the possibility of a single answer range from zero to $\pm\infty$; it seems impossible. I tried to research the problem, finding there was nothing on the concept! It wasn't until I noticed a square map of the world on the wall of my teacher's room that I had an idea, however it wasn't regular mathematics.

Say a person is standing at the center of a square map (I realize it's usually the middle of the Atlantic Ocean, it's just an example), and this person wants to go to the south-most part of the world. If we look at this square map as a grid it appears as though there's a similar problem, there are infinite ways to get to the South Pole (or the limit) which appears as a line when it's truly a single point. With this logic it's reasonable that the entire limit of the x, y grid could be a single point, and the best way to visualize this is as the surface area of a sphere.

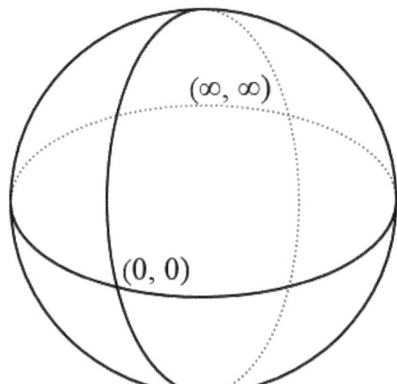

This is an abstract idea: a three-dimensional way to portray a two-dimensional grid. However we have been using square units in a spherical universe, and we've used it effectively. Imagine the possibilities if we used a grid that better fit this mold.

Chapter4: Discoveries of the Spherical Grid

When trying to figure out the spherical grid, some advantages and disadvantages arose for each method that was tried to effectively graph infinity. Before I get into the different types of spherical grids I need to first go over some of the basic details. First of all, the units on the x-axis and y-axis are evenly spaced evenly along the circumference from ±0 to ±∞. Then each of the quadrants is placed in the same correlation of the origin.

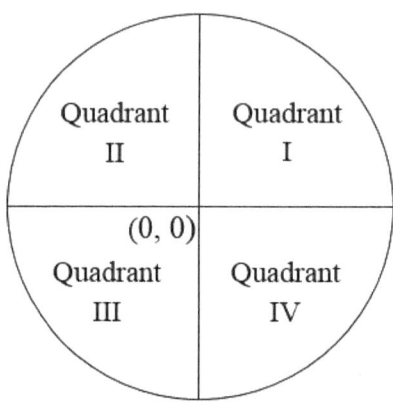

Front View

One day I was wondering why we struggle so hard to make our understanding of the universe fit the math we use. If we adapted the math we use to help fit the problems we faced then our own discoveries would improve with our tool to discover. Someone reading this book may ask how the units warp between other units to fit the sphere. My answer is several methods, as shown below, to fit different scenarios both physically and geographically.

1. Cartesian 2D Grid (.1) – This is the square-based grid we typically use. There are some problems with regular graphs to ∞ with this method.

2. Balanced Side Spherical Grid (.2) – Every unit is composed of circumferences that revolve around the grid on the points halfway on the x-axis and y-axis. A key point to remember is that every unit has a circle that connects it to the unit on the opposite side of the sphere. Because of this principal the graphs x=∞/2 and y=∞/2 are the exact same, this is the midsphere line (I know it's really a circle, though it's still a linear graph).

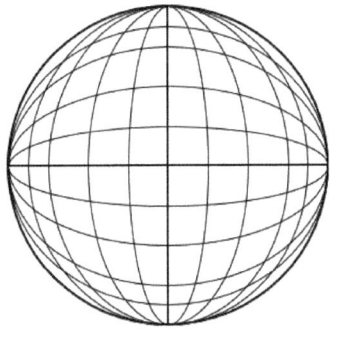

Front view

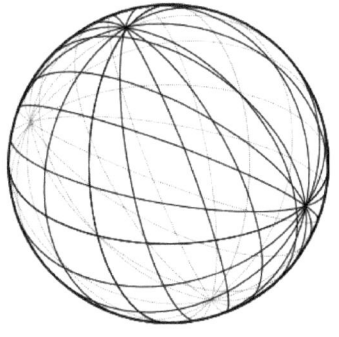

Angled view

3. <u>Equal Side Spherical Grid</u> (.3) – With this method the units lead to circles parallel to the x-axis and y-axis. Keep in mind that the flaw in this method is that the graphs x= ∞/2;   x= -∞/2; y= ∞/2; and y= -∞/2 are only points on this grid.

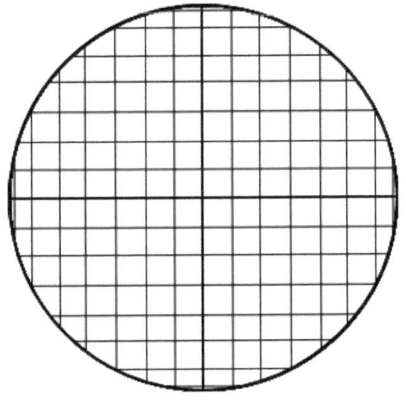

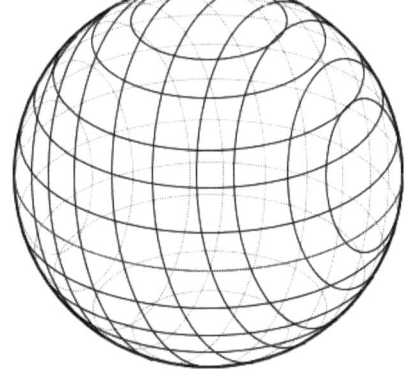

Front view

Angled view

4. <u>Singularity Spherical Grid</u> (.4) – This format best portrays the Cartesian grid warped into a sphere. For this grid method, every unit forms a circle that must pass through the limit point (∞, ∞) once. Imagine a regular 2D Cartesian grid where (0, 0) on it meets (∞,∞) on the spherical grid, this is the <u>slope plane</u> and only works for this grid. If you then create a line that is tangent to the linear graph on the sphere that lies on the slope plane you have a <u>slope tangent</u>, which translates to the slope plane and defines the slope of the graph on the spherical grid. Graphs with the same slopes have the same slope tangents.

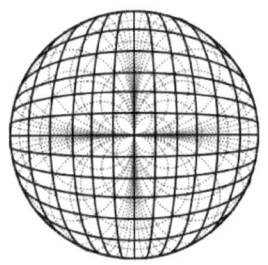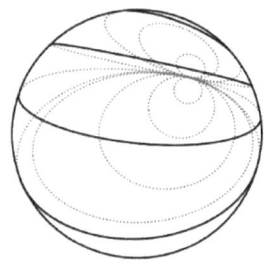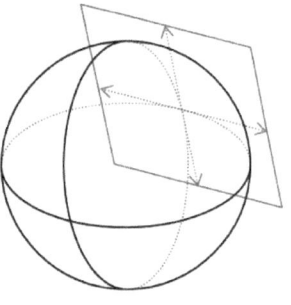

| Front view | Angled view (y-units shown only) | Slope Plane |

If you look back on the definitions of the types of grids you'll notice there are numbers in blue, these are <u>indication numbers</u> to tell which grid to use for a graph: the Cartesian 2D grid is .1, the balanced side spherical grid is .2, the equal side spherical grid is .3, and the singularity spherical grid is .4. To distinguish, put the full equation in indication brackets (what I'm calling the symbols ʕ ʔ), then give the brackets a base of the indication number to show which grid to use.

Example: ʕy=xʔ.2 Is graphed on the balanced side spherical grid.

Chapter5: Values of Infinity, Morphed Grids, and Between Value Chains

The symbol to give a value to infinity is ∞_v. As stated before, infinity can be a limit, not always an unreachable value (the default value of infinity is unlimited if not specified, or $\infty_v=\infty$). When a value is given to infinity there are ∞_v unit spaces from 0 to ∞ and ∞_v decimal spaces between each unit (∞ and $1/\infty$). Thus every positive number on an axis from 0 to ∞ when $\infty_v=3$ is 0, 1/3, 2/3, 1, 4/3, 5/3, 2, 7/3, 8/3, and ∞. There can be nothing else on the grid because nothing else can exist on the grid.

Example: The graph y=4x can't exist on a grid where $\infty_v=3$. Because there is no ¼, ½, or ¾ on the grid and if it is still graphed is classified as <u>irregularly existent</u> on this grid.

To give the same value to infinity on every axis you give a value to ∞_v, however to give separate values to each axis you use ∞_x, ∞_y, and ∞_z. Giving different values to infinity on each axis allows more flexibility when graphing.

There are also <u>between value chains</u>, which allow value fluctuations between 0 and ∞ along a single axis to easily portray multiple orbits on the grid. The most basic is a 0-∞-0 chain, which is when there are two point (0, 0)s opposite to each other on the sphere with infinity value between them. The default is that infinity is exactly between the two points. However (as an example) to edit this, have in regular parenthesis before the identification brackets (0-∞-0; v=10%), this means infinity is ten percent of the axis's total distance from the origin (0, 0) and ninety percent of the axis's distance from the limit (0, 0). Keep in mind there are the same number of units between the (0, 0)s and infinity. Also note you can have multiple of these between limits, such as 0-

∞-0-∞ etc.; the only rule is the numbers must alternate, then to give their distances from the front value you use percentages: (0-∞-0-∞; v=50%, 60%). To keep it simple, each percentage must be in increasing order, the first infinity here is 50% of the axis distance while the second zero is 60% of the total axis distance from the origin. Then if the scenario required you to give individual between value chains to each axis you would have multiple parenthesis replacing v with x, y, and z.

Example format: ∞_v=100; (0-∞) (0-∞-0-∞; y=30%, 80%, 85%); ⌠x=y⌡$_{.2}$

Note: The x-value for the between value chain (first chain) isn't specified because it is at its default

   With all this talk of individual axis values you may be wondering about the indication number. The way the indication number works is if you have one, it applies to every axis; if you have multiple numbers then each affects the axis in the order it's written in (x, y, z order, which is vital to every part of formatting in Sphereology).

Example: ⌠x=y⌡$_{.2, .3}$; This means the x-axis's units follow the balanced side spherical grid and the y-axis units follow an equal side spherical grid format, this is a <u>morphed grid</u> and this example is how we format longitude and latitude for our world.

Notes: Sphere grids **<u>won't</u>** morph with cubic grids, thus ⌠x=y⌡$_{.1, .3}$ is impossible. Also keep in mind that any value plus ∞_v is the value opposite on an axis, while a value plus $2\infty_v$ becomes the exact same value on any grid.

Chapter6: The Third Dimension

The surface area of a sphere is two-dimensional, to get depth you must create layers of surface areas. The z-axis is unique in several ways, first is its different axial path from the x-axis and y-axis, it is visually linear instead of circular. Plus this is the only axis where infinity and negative infinity are visibly separate (though are truly equal). Infinity on the z-axis is the outermost layer of the sphere while negative infinity is a single point at the sphere's center, leaving the zero layer between them.

Note: All two-dimensional layers on the same grid must have identical properties.

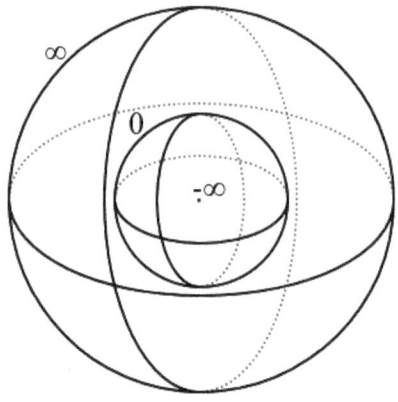

Z-axis layers

This format gives the z-axis unique between value chains, usually the minimum value chain is 0-∞, for the z-axis the minimum is reciprocated to ∞-0-(-∞). With the x-axis and y-axis the pattern on the between value chain was 0, ∞, 0, ∞, 0, ∞, etc.; on the z-axis (and any axis on a Cartesian grid) the chain pattern is ∞, 0, -∞, 0, ∞, 0, -∞, 0, etc. This chain pattern is because 0 is always between positive and negative infinity.

On the spherical 2d grid the point (∞, ∞) is identical to the point (-∞, -∞) the same as the point (0, 0) is the same as (-0, -0) on any grid, thus ∞ is equal to -∞ depending on correlation to infinity (Thus all the between value chains are in a way identical). For the z-axis this means the entire grid must be within itself. The following are the different ways graphs can react to infinity on the z-axis.

1. Cartesian 3d Grid (.1) – A cubic Cartesian grid with limits. The between value chain pattern for all square-based grids is the same as the z-axis on the spherical grid: ∞, 0, -∞, 0, ∞, etc.

2. Balanced Side z-axis (.2) – When a graph goes to infinity or negative infinity on the z-axis it continues on the opposite limit from the opposite side of the sphere.

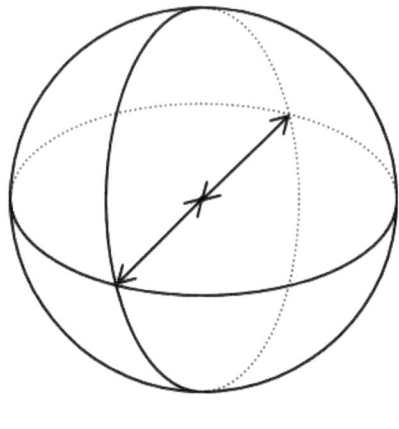

x=y=0

3. Equal Side z-axis (.3) – When a graph goes to infinity or negative infinity on the z-axis it continues on the opposite limit from the same side.

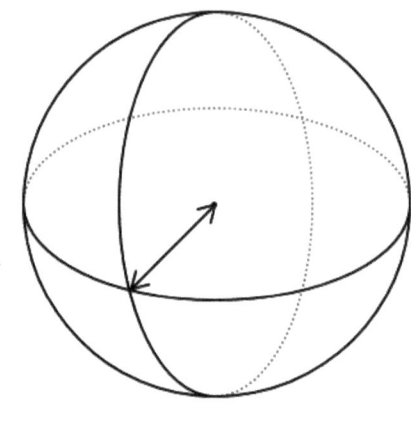

$\underline{x=y=0}$

4. Passive Singularity (.4) – If a graph goes to negative infinity on the z-axis it will "just pass" the point and continue onward until it reaches the infinity layer, then ends.

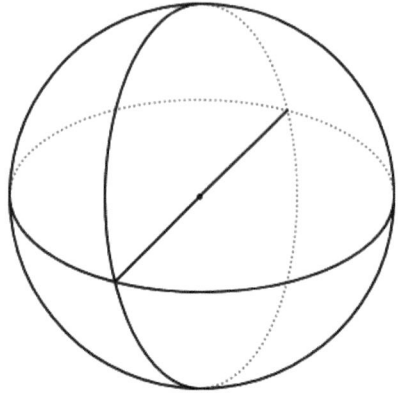

x=y=0

If the new age of math is meant to be as flexible as our imagination then should we be limited to evenly-spaced units? After some thought there could be some uses to differently spaced units, such as to graph falling objects that exponentially accelerate to a source of gravity or to model relativity (changing the grid instead of the equation). Before the parentheses, have in brackets v equals a fraction, decimal, or percent. What this means is between zero and ±infinity every unit space is the percent *size* given of the previous unit. Say a unit space is 10% of the entire axis, and inside the brackets is [v=40%], the next unit would be 40% the size, thus would take up 4% of the whole axis.

Note: The space a unit takes up is not based on volume or area, it's based on the linear value of the axis.

At the base of the bracket put either 1 or -1, one means from zero to infinity (and from the zero layer outward on the z-axis or cubic grid), and negative one is vice versa. The <u>initial value</u> is the percentage the first unit space takes up on the axis; this cannot be any random number because when every percentage of every unit space is added up from the origin to the limit it must equal 100% of that half of the axis. To separate the axis values you would have three separate brackets, replacing v with x, y, and z.

Example: $\infty_v = \infty$; $[v=10\%]_1$; On this grid the initial values on the positive and negative x-axis are the unit spaces between 0 and ±1 due to the 'forward' correlation. With this percent, the initial values must be 90% of the entire axis because then the next unit spaces would be 9%, then 0.9%, 0.09%, … until the units on both sides of the axis add to 100% and reach infinity. If in the brackets was $[v=10\%]_{-1}$ then the unit spaces between ∞ and $\infty\pm1$ would take up 90% of the axis.

Note: If $[v=1]$ then you don't need a base of 1 or -1 because every unit is identical.

    Another method for exponential units is to use an equation or multiple equations in the brackets. How this method works is each unit's absolute value is plugged into the equation, the outcome of the equation is the unit's <u>placement value</u>, when every unit's placement value is added up the outcome is the <u>axis value</u>. To find where to place each unit (as a percent decimal from other units) divide the placement value by the axis value.

    One potential flaw (and benefit) from exponential units is with between value chains altering the units to decrease, then increase again. Should one wish to have a steady decrease or increase in exponential units from the origin to the limit despite the chain they would use a <u>constant unit change</u>. To utilize this constant you replace the regular brackets [ ] that hold the exponential unit and replace it with { }. The squiggly brackets treat the relation between the origin and its opposite part as it would be to its standard value chain. Thus $\infty_v=5$; $\{x=10\%\}_1$ (0-∞-0-∞-0) would have the first initial values of 90% and would steadily decrease without increasing again despite the between value chain.

Chapter8: Value Placement or Rise for Run

    On a standard Cartesian grid the placement of a point and the "rise for run" method of graphing go side by side. On a spherical grid however there grows a distance between the methods. The default we will go by is placement of a point based on the rise for run method. However you could underline the part of the equation in indication brackets that will rely on value placement over this method.

Example: $\{x=y^2; \underline{x=3y/7}\}$: The x=3y/7 graph relies on value-based placement while $x=y^2$ will follow the rise for run rule, ignoring placement by value.

Chapter9: Zero and Infinite Kingdoms and the Infinity is One
Principal

If you've tried spherical graphing for practice with an exponent
graph you probably realize the disaster of it (the graph spirals the
sphere in different ways depending on the value of infinity). To
overcome this abstraction and make problems like the exponent
regular as we would prefer, we use 0 and ∞ *kingdoms*. How 0 and
∞ kingdoms work is by first dividing the grid in half between the
origin and its limit point into two hemispheres or *kingdoms,* the
half towards ∞ being the infinite kingdom. The zero kingdom
follows the rules of everything we have discussed thus far while
the infinite kingdom follows the inverse properties of its opposite,
thus making infinity a second origin and making the kingdoms
mirrored versions of each other. To have part of a graph follow the
kingdoms rule overline the equation in indication brackets. Even
then this graph can look strange depending on the value of infinity,
to make this graph elliptical along the surface area you also put the
desired part of the equation overlined in between two asterisks
(*x=y*) to apply the infinity is one principal. What this principal
by itself does is turn the units into parts of the value of infinity and
treats infinity as one, if infinity were twelve then the units become
twelfths of the axis and the decimals become twelfths of the
twelfths. Continuing with infinity as twelve the graph for $y=x^2$,
when x is six it is actually $6/12$: $(6/12)^2 = 36/144 = 3/12$, thus the y-
value when x is six is three. Without the kingdoms rule this
principal returns the example back to the origin without reaching
infinity, with both the square exponent becomes an ellipse along
the grid

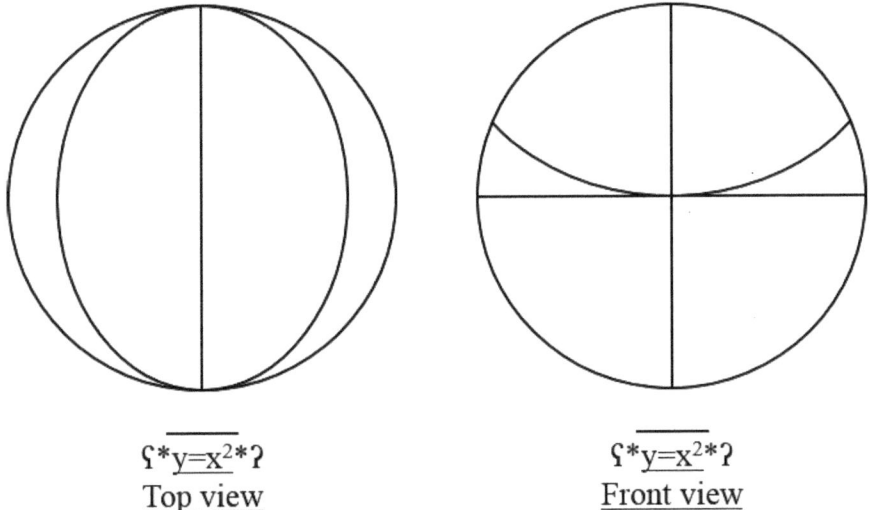

$\varsigma * \overline{y=x^2} * \varsigma$
Top view

$\varsigma * \overline{y=x^2} * \varsigma$
Front view

<u>Complete Format and Summary of Three-dimensional Grids</u>

Example: $\infty_{x,\,y}$=40, ∞_z=20; [x=1] {y=10b}$_{-1}$[z=67%]$_{-1}$; (0-∞-0-∞; x, y= 50%, 60%); ∫*y=3x+7*; z=4x/y?$_{.4,\,.4,\,.2}$

1. In <u>Red</u>: Given values to infinity for each axis:

   a. Instead of writing ∞_x=40 and ∞_y=40 you can group bases together

   b. V is equal to x, y and z.

   c. You only specify an axis if it is in use.

2. In <u>Orange</u>: Exponential units:

   a. When x=1 then all units on the x-axis are evenly spaced

   b. At the bases of each bracket is a one or negative one (if applies), which describes the correlation of the percentage or equation as forward (the origin to the limit) or backward (vice versa).

   c. A percentage means the next unit space is the given percentage of the one previous to it.

   d. In an equation, if x, y, or z represents the placement value of the axis, then there can only be one other variable representing the value of the unit, usually a, b, or c.

   e. To find where to linearly place each unit with an equation (as a decimal-percent from other units) divide the placement value by the axis value (all placement values of one side of an axis added together).

f. The squiggly brackets ignores the between value chain of the axis for an equation and treats it as one continuous axis: for the y-axis in this example the constant unit change treats the axis as it having 120 units on each side with 40 decimal places as the exponential units apply while still following the percent placements of the extra values.

g. The placement value for a unit on the y-axis is not related to its unit value after the first infinity due to the constant unit change.

3. In Green: Between value chains:

a. These chains are shown if there is more than the default chain for an axis, never less.

b. The chains are in order from the origin to the limit (the starting point to its opposite).

c. Only because the x-axis and y-axis have identical between value chains they were grouped together the same way as the values given to infinity.

d. Each between value chain for an axis applies to both sides of the axis and show the values from the origin to the limit, for the x and y axes there are then three zeroes and three infinities along the entire axes.

4. In Blue: Indication brackets with the equation to graph:

a. The underline means the graph relies on value placement over the rise for run method.

b. An overline (not shown in example) means that graph follow the zero and infinity kingdoms rule, which is a property where the two kingdoms mirror each other by giving inversed properties.

c. With the z-axis, the axes on a Cartesian grid, and with between value chains there can be multiple kingdoms so long as the boarder is exactly between infinity and zero.

d. The equation between the asterisks means the graph follows the infinity is one principal.

5. In Purple: Identification numbers in x, y, z order:

a. If there is only one identification number then that number applies to all axes, otherwise every axis must be identified.

b. A grid can either be spherical or cubic, never both.